SO-ARL-653

The Science of
RENEWABLE
ENERGY

THE SCIENCE OF
HYDRO *AND*
WAVE ENERGY

by James Bow

MATTESON AREA PUBLIC LIBRARY
DISTRICT

ReferencePoint
Press®

San Diego, CA

© 2018 ReferencePoint Press, Inc.
Printed in the United States

For more information, contact:
ReferencePoint Press, Inc.
PO Box 27779
San Diego, CA 92198
www.ReferencePointPress.com

ALL RIGHTS RESERVED.

No part of this work covered by the copyright hereon may be reproduced or used in any form or
by any means—graphic, electronic, or mechanical, including photocopying, recording, taping,
web distribution, or information storage retrieval systems—without the written permission of
the publisher.

Library of Congress Cataloging-in-Publication Data

Names: Bow, James, author.
Title: The science of hydro and wave energy / by James Bow.
Other titles: Science of hydro and wave energy
Description: San Diego, CA : ReferencePoint Press, Inc., [2018] | Series: The
    science of renewable energy | Includes bibliographical references and
    index.
Identifiers: LCCN 2017043216| ISBN 9781682823033 (hardcover : alk. paper) |
    ISBN 9781682823040 (pdf)
Subjects: LCSH: Water-power--Popular works. | Ocean wave power--Popular
    works. | Hydroelectric power plants--Popular works.
Classification: LCC TJ840 .B685 2018 | DDC 621.31/2134--dc23
LC record available at https://lccn.loc.gov/2017043216

IMPORTANT EVENTS IN THE DEVELOPMENT OF
HYDRO AND WAVE ENERGY

1822 CE
British scientist Michael Faraday produces the first electric motor.

1880
Hydro power lights a theater and storefront in Grand Rapids, Michigan.

250 BCE
The first water-powered wheel is invented in Perachora, Greece.

300 BCE	200 BCE	100 BCE	1850 CE	1880

200 BCE
Water power is used to power simple machinery by the Han Dynasty in China.

1878
In the country estate of Cragside in Northumberland, England, a waterwheel is used to turn a turbine, powering a dynamo to generate electricity for the first time.

71 BCE
Water is used to power a grain mill at Cabira, Asia Minor.

1895
The Adams No.1 generating station at Niagara Falls begins supplying electric power to nearby industries.

4

1920
The US Army Corps of Engineers is authorized by the US government to build hydroelectric plants.

2016
Nova Scotia Power and Cape Sharp Tidal test an experimental turbine to generate electricity from the Bay of Fundy, Nova Scotia.

1973
The oil crisis, started by an oil embargo by oil-producing countries in the Middle East against the United States, raises oil prices dramatically, sparking interest in renewable energy sources.

1966
France opens the first tidal power generating station in the Rance River Estuary.

1984
The Itaipu Dam opens, straddling the Paraná River between Brazil and Paraguay.

1920　**1940**　**1960**　**1980**　**2017**

1936
Hoover Dam is built on the Colorado River on the border of Arizona and Nevada, creating Lake Mead.

2008
Portugal opens the Aguçadoura Wave Farm, the first experimental power plant generating electricity from ocean waves.

1929
The first large-scale pumped storage plant is installed on the Housatonic River in Connecticut.

ENERGY
FROM *WATER*

FROM THEORY TO APPLICATION

Moving things have kinetic energy. When water is in motion, whether in a rushing river or in the tides and waves of the ocean, its kinetic energy can be harnessed by mechanical devices. In rivers, gravity pulls water downhill. Hydroelectric power plants take advantage of this motion, using the moving water to spin turbines. These turbines are connected to generators that turn kinetic energy into electricity. In the ocean, the gravity of the Sun and moon create tides. Wind creates waves on the water's surface. And underwater, differences in temperature create ocean currents. Engineers have created various devices to turn the motion of tides, waves, and currents into electricity. These power sources produce far less emissions than burning fossil fuels, and they are renewable resources. However, hydroelectric power may not easily replace fossil fuels in all situations, and these power stations have environmental problems of their own. Ocean energy sources present major technical hurdles to overcome. Dealing with these issues will be the challenge for scientists over the next few years in the quest for a more environmentally friendly future.

Niagara Falls is a scenic destination for millions of visitors each year. It is also a source of electricity for the United States and Canada.

Millions of visitors come to see Niagara Falls each year. They stand in awe, witnessing the raw power and thundering sound of thousands of tons of water cascading 176 feet (54 m) down a sheer cliff. More than 200,000 cubic feet (5,660 cubic m) of water flow through the Niagara River every second.

But visitors may be surprised to learn that they are seeing only a portion of the river's water. Niagara Falls is both a natural wonder and a power source, providing enough energy for more than 3 million homes. Power plants on both sides of the Niagara River divert most

WORDS IN CONTEXT

megawatt
One million watts.

erosion
The process by which rock is worn away over time by the movement of water on it.

of its water through tunnels and turbines to generate more than 4,400 **megawatts** (MW) of electricity. These tunnels and channels can divert so much water from the river that Niagara Falls can run dry. The Army Corps of Engineers carried out such a complete diversion to the American falls in June 1969 to do a geological survey of the rock face of the falls, measuring **erosion**. Treaties signed by the Canadian and American governments require that the power plants on the Niagara River allow at least 100,000 cubic feet (2,830 cubic m) per second to pass over the falls during tourist season.

More than 600 miles (970 km) to the east, in Canada's Bay of Fundy, electricity has been generated from the movement of water in a different way. As the high tide raises the level of the Atlantic Ocean, water is funneled into the narrow bay. With nowhere to go, seawater piles up, raising water levels by as much as 50 feet (15 m). On November 22, 2016, a five-story-tall, 1,100-ton (1,000-metric-t) turbine was lowered into the water in an area called the Minas Passage. It was then connected to the power grid. Looking like a massive doughnut, the device let the waters of the Bay of Fundy pass through, turning blades and generating enough electricity to power five hundred homes.

Matt Lumley of the Fundy Ocean Research Centre for Energy described how the potential electricity generation exceeded expectations: "Initial estimates for the potential of the Minas Passage site put it at around 300 megawatts, about 10 per cent of Nova Scotia's peak electricity demand. But once we actually got into the bay and started to collect some field data . . . that number went up significantly to about 7,000 megawatts of power."[1]

The device, made by Cape Sharp Tidal, successfully generated electricity from November 2016 to April 2017, before being retrieved at the end of the test. Additional tests were planned to further refine the technology. There are hopes that, if these tests are successful, the Bay of Fundy could generate enough electricity to power many thousands of homes in eastern Canada.

Finally, near the Rock of Gibraltar, located where Spain nearly touches Africa, a set of experimental buoys have been laid out on the sea to capture the rise and fall of the waves themselves. The planners of the project, which is part of an agreement made between Eco Wave Power and the government of Gibraltar, hope that these devices can generate 5 MW of electricity by the completion of the project, projected for 2020. While small compared with other power plants, this is enough to supply 15 percent of Gibraltar's power needs. Journalist Chris Wood, who attended the project's opening ceremony, noted that it was a strong positive sign for future development: "Overall, the low-impact nature of the energy solution, combined with its easy-access maintenance and long life span could allow it to have a big impact across the globe."[2]

The State of Hydro and Wave Power Today

Hydro power, the harnessing of the energy of moving water, is both an old renewable resource and a modern one. People have been using the force of waterfalls and moving streams for millennia. Today, nine out of the ten largest power plants in the world run on hydro power from dammed rivers. Their combined generating capacity tops 95,000 MW. There are many benefits to hydro power. Once the dams are built, the energy is cheap to produce and generates far less **pollution** than burning fossil fuels.

WORDS IN CONTEXT

pollution
Waste products that change the air, soil, or water in ways that are harmful to life.

Hydro power is not perfect, however. There are ecological consequences to damming rivers and running water through hydroelectric power plants. Builders of hydroelectric plants have to balance the needs of the environment and the needs of people living near dams with the benefits that the power plants provide. Failing to do this hurts wildlife, displaces people, and even contributes to global climate change.

Engineers have gained a great deal of experience at converting the kinetic energy of rivers into electricity, but this represents only a fraction of the world's moving water. The ocean offers tides that rise and fall, surface waves generated by the kinetic energy of wind, and underwater flows of water known as currents. Researchers are

The world's first large-scale tidal power plant, located on the Rance River in Brittany, France, still operates today. Its twenty-four turbines generate a maximum of 240 MW.

only beginning to tap the potential of these energy sources. The first large-scale tidal power plant was built in France in 1966, but technical challenges and environmental concerns have limited the spread of such systems. Fewer than a dozen of these plants operate today. However, new turbine designs, such as the one tested in the Bay of Fundy in 2016, could make it easier to capture the energy of the tides with less environmental impact. The makers of the largest tidal power station yet conceived, the MeyGen Tidal Energy Project, hope to eventually generate 398 MW of power from the waters of the Pentland Firth in northern Scotland. The first of four underwater turbines started working in December 2016. Nicola Sturgeon, the first minister of Scotland, hailed the project as part of the fight against environmental damage and climate change: "I am incredibly proud of Scotland's role in leading the way in tackling climate change and investment in marine renewables is a hugely important part of this."[3]

Technology to generate power from the movement of waves or ocean currents is even more experimental. The ocean is challenging to work with, and sites are often far from the electrical grid. However, engineers are designing and testing new equipment. While an attempt to generate 2.5 MW of power offshore from Portugal in 2008 failed, companies in England, Scotland, and the United States are pursuing other designs in hopes of finding a breakthrough. Small-scale, experimental power plants are already operating off the coast of Scotland, while similar projects are planned in Mexico and China.

Flowing Forward

Tapping into the full potential of hydro and wave power would make it possible for us to replace some of the electricity now generated by burning fossil fuels. Sam O'Neil, the president of the Ocean Renewable Energy Coalition, said, "The total potential off the coast of United States is 252 million megawatt hours a year. That's equal to about six-and-a-half percent of our total capacity in the United States, equal to all the dams that we have in the US right now."[4]

Challenges remain, however, including how to use hydro and wave power in ways that are currently more suited to fossil fuels, such as powering motor vehicles. There are also the environmental and social impacts of large-scale hydro projects to consider. The dams used to create reservoirs for hydroelectric power plants have significant issues. The Union of Concerned Scientists explains, "Flooding land for a hydroelectric reservoir has an extreme environmental impact: it destroys forest, wildlife habitat, agricultural land, and scenic lands. In many instances, such as the Three Gorges Dam in China, entire communities have also had to be relocated to make way for reservoirs."[5] Other elements of hydro and wave energy have problems, too.

No renewable energy source is without its drawbacks. But while hydro and wave energy sources are unlikely to completely replace fossil fuels on their own, they can work alongside a mix of other renewable sources, such as wind and solar power, to reduce reliance on dirty and often dangerous fossil fuels. Engineers and scientists are now focusing on how to improve hydro and wave power technology, boosting **efficiency** and power output while reducing costs and environmental risks. These energy sources are likely to be a key part of our renewable energy future.

WORDS IN CONTEXT

efficiency
The ability to capture more energy from a given energy source.

HOW DO HYDRO AND
WAVE ENERGY *WORK?*

In the third century BCE, inventors in Greece and in Byzantium placed wheels with paddles on them along flowing streams. The flowing water pushed against the paddles, causing the wheels to rotate. Ropes and pulleys transferred this motion to tools that could grind grain into flour. As time went on, waterwheel designs improved. Early wheels were laid on their sides in the current. These were replaced with stream wheels, where the axis was horizontal, and the bottom of the wheel sat in and was pushed by the stream. Since most of the wheel was outside the stream, where air offers less resistance than water, the wheel could turn faster, using less water. Inventors found ways to use the wheels' motion to saw logs and pump bellows to stoke furnace fires.

The Chinese also learned how to use water to do work around this time. During the Han Dynasty, after 202 BCE, the Chinese used waterwheels to power hammers that broke rocks to extract ore or pounded wood pulp to make paper. Around 10 BCE, Roman inventors

Waterwheels are still in use in many rural areas of the planet. This one is used to irrigate fields in Cambodia.

started using drains or sluices to direct the water onto the wheel from above. Gravity pulling on the water would cause it to impart more force to the wheel, spinning it more strongly. As the Europeans and the Chinese used and experimented with their technology, they continually made improvements so that the water could do more work. Still, the popularity of waterwheels remained limited due to the availability of other energy sources, as engineer B. J. Lewis explains: "The availability of slave and animal labor restricted [water power's] widespread application until the 12th century."[6]

The Science of Water Energy

Water is a fluid. This means it deforms and flows when a force is applied to it. As gravity pulls it downhill, it gathers momentum.

Its kinetic energy increases. When water hits the blade of a waterwheel, some of that kinetic energy is transferred to the wheel as the wheel goes into motion.

The kinetic energy of something is calculated by multiplying the mass times its velocity squared. Water can have a great deal of kinetic energy, because it is heavy—more than eight hundred times denser than air. A cubic meter of air near sea level weighs about 2.6 pounds (1.2 kg), whereas a cubic meter of water weighs about 2,205 pounds (1,000 kg).

The other ingredient in kinetic energy is velocity. For water to have kinetic energy, it must be moving. Water that naturally flows downhill is one easy source of moving water. Streams and rivers work well, but waterfalls are the real powerhouses. The bigger the change in height, the faster gravity can get the water moving. The faster the water's speed, the greater its kinetic energy is.

Moving Magnets to Create Electricity

By 1750, people had been using the kinetic energy of water to do work for two millennia. But waterwheels had to be located near water, so the power they generated could only be used in limited areas. A huge technological leap forward was needed to transfer the power generated from streams and rivers farther away.

Between 1600 and 1750, scientists such as William Gilbert, Otto von Guericke, Robert Boyle, and Benjamin Franklin started to explore the phenomena of electricity and magnetism. British scientist Michael

Faraday made a great breakthrough in 1822. Months before, scientists had shown that when electricity was sent through a wire, the wire became magnetic. To learn more, Faraday made a device that dipped a wire in a pool of the metal mercury. When Faraday sent electricity through the wire, the wire moved in a circle through the mercury. Faraday had created the first crude electric motor, turning electrical energy into motion. Further tests showed that this process worked in reverse, too. Moving a magnet through a coil of wire could produce electricity. This is the basic principle behind a generator.

Faraday slowly worked out the relationship between electricity and magnetism. He learned that a group of moving electrons, or an electric current, creates a magnetic field. Moving magnets create an electric current. At the time, some considered electricity to be nothing more than a curiosity, but Faraday and other scientists knew it could be useful. When a British government official asked Faraday what electricity was good for, Faraday shot back, "Why, sir, there is every probability that you will soon be able to tax it!"[7]

Water was an obvious choice to move magnets to generate electricity. A device that does this is called a turbine. Benoit Fourneyron, a French engineer, invented one of the first turbines in 1827. James B. Francis, a British-American inventor, improved the design for a water-powered textile factory in Lowell, Massachusetts, a few decades later. These high-efficiency turbines meant that a lot of electricity could be generated from even small flows of water. Francis's work was highly influential. He published the results of his experiments in a book called *Lowell Hydraulic Experiments*, which the American Society of Civil Engineers has described as "a landmark text."[8] At the same time, other inventors were creating devices that used electricity. Scotsman James Bowman Lindsay invented an electric light bulb in 1835. The telegraph, invented in 1816 by Englishman Francis Reynolds, was perfected in 1837 by William Fothergill Cooke and Charles Wheatstone.

The two sides of the equation—power generation and electricity use—were brought together in 1870 in a country house called Cragside in Northumberland, England. Lord Armstrong built a waterwheel by one of his estate's lakes and attached it to a generator, which he used to power an arc lamp, making Cragside the world's first hydroelectric power station. Thanks to Armstrong's high-tech work, the home became known as the "palace of the modern magician."[9]

By the 1880s, inventors including Thomas Edison were setting up electrical grids to power lights and electric trams throughout cities. Edison was quoted in the *New York Herald* as saying, "After the electric light goes into general use, none but the extravagant will burn . . . candles."[10] Much of this early electricity came from hydroelectric power plants. There are still people, including those in Ontario and New York, near Niagara Falls who refer to their electrical power as "the hydro," even when it comes from other sources.

Using Dams to Increase Water's Potential Energy

Many hydroelectric plants are near waterfalls. One of the earliest was built in 1882 at Niagara Falls. Here, the Niagara River flows from Lake Erie to Lake Ontario. The water's elevation drops 330 feet (100 m) from lake to lake. Today, the Sir Adam Beck Pump Generating Station generates up to 174 MW of energy by diverting water from the Niagara River through tunnels paralleling the river.

There are a limited number of waterfalls in the world, so engineers sometimes make their own. To do this, they use hydroelectric dams. A dam blocks the flow of water, creating an artificial lake behind the dam. A pipeline or a sluice sends some of that water roaring through turbines. Essentially, hydroelectric dams use the weight of the lake to create the kinetic energy to turn the turbines. Squeezing that water through small pipes adds to the force that is placed on the turbines.

A big dam can be even more powerful than a big waterfall. For example, Hoover Dam holds back the Colorado River and creates Lake Mead. The flow of the Colorado itself is not predictable enough or forceful enough to generate a lot of electricity, but the weight of the water behind the dam can generate up to 2,000 MW through the dam's turbines when the lake is full. The dam is enormous. As author Michael Hiltzik notes, "This wasn't only the biggest dam in the world at the time; it was more than three times the size of the biggest dam that had ever been built."[11]

19

Lake Mead is 112 miles (180 km) long and a maximum of 532 feet (162 m) deep. That depth is important. Whether the hydroelectric plant is at a dam or a waterfall, the distance that water falls is called the head. The higher the head, the greater the potential energy of the water. As the water falls, this potential energy becomes kinetic energy—the energy of movement.

Types of Turbines

To turn the kinetic energy of water into electrical power, water must push a turbine, which spins a generator creating electricity. Hydropower turbines are sorted depending on how the water moves

Calculating the Energy of a Hydro Power Plant

The energy produced by a hydroelectric power plant depends on the head, the flow, the force of gravity, and how efficient the turbines are in changing kinetic energy into electric energy. The head is a measure of the height of the waterfall, or the depth of the dam. The flow is a measure of how much water is moving through the generator per second. The gravity is the only constant in the calculation. All objects on Earth, when dropped, accelerate by 32.2 feet per second (9.81 m/s) every second the object drops. This includes water.

Multiplying the head by the flow and the acceleration due to gravity gives the total power of the water. For example, if the head is 39 feet (12 m), and the river flows at 7 cubic feet (0.2 cubic m) per second, then with gravity, the total power of the river is 23,500 watts of energy, or 23.5 kilowatts.

No turbine is 100 percent efficient in changing the motion of water into electricity. Many are 60 to 70 percent efficient. At 70 percent efficiency, a turbine on that same river would generate 16.5 kilowatts of power.

them, and whether they're mounted vertically (turning like a hamster wheel), or horizontally (turning like a spinning plate).

The two main types are called impulse and reaction. Impulse turbines rely solely on the force of gravity as water falls against the blades or buckets, pushing them down and causing the wheel to turn. Reaction turbines rely on differences in the water pressure on each side of a blade in order to turn the wheel.

One issue that turbine designers face is friction. Friction occurs when materials move against one another. Friction acts to resist that motion. Think of how much energy you need to drag a box across a floor, and how much easier it is to pull that same box over the same floor if it is mounted on wheels. Wheels generate less friction, and so the box is easier to move.

Friction is almost impossible to eliminate, and it reduces the efficiency of transferring kinetic energy from the water to the turbine. The wasted energy becomes heat. The more friction is reduced, whether through better lubricants, ball bearings, or other designs that limit the amount of contact surfaces have against each other, the more electricity a turbine can generate from the same amount of kinetic energy.

Power from the Tides

Rivers and lakes hold only a fraction of Earth's water. Oceans cover more than 70 percent of the planet. Because the ocean is relatively flat, electrical power generated from the ocean can't rely on downhill flows. Instead, it comes from the ocean's tides, waves, and currents.

Although the ocean is roughly flat, it isn't still. Consider the tides. Earth's gravity is constantly pulling water toward the planet's center, but another force is working against that pull. The gravity of the moon pulls on Earth and the water on the planet's surface. It creates a bulge in the water closest to the moon. A bulge of water exists on the side of Earth opposite the moon as well, caused by the moon pulling Earth away from the water on that side.

Earth turns beneath the moon, and the moon orbits Earth. These two motions pull the tidal bulges across the planet. The result is that each spot on the planet sees a high tide roughly every twelve hours and twenty-five minutes. The Sun's gravity also pulls on Earth's surface, creating its own tides, but these are smaller and slower. When the moon and the Sun are aligned, however, especially when Earth is the closest to the Sun in its orbit, their forces can combine to create even larger tides.

The tidal difference is not very noticeable in small bodies of water or in the middle of the ocean. Even something as large as the Mediterranean Sea or the Baltic Sea rises and falls only about 1 foot (30 cm) between high tide and low tide. However, in some places, the shape of the coast funnels the tides through smaller spaces where the tidal differences can be much larger. The Bay of Fundy sees water levels change by as much as 52 feet (16 m) between low tide and high tide. The government of Canada describes it as having "the highest

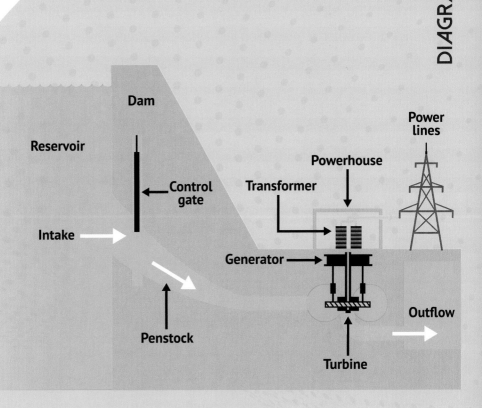

Dam

Reservoir

Control gate

Power lines

Powerhouse

Transformer

Intake

Generator

Penstock

Outflow

Turbine

Inside a Hydroelectric Dam

The inner workings of a hydroelectric dam are relatively straightforward, but the structures themselves can be massive in scale. Huge masses of concrete are often necessary to hold back reservoirs that become large artificial lakes. Gigantic turbines and generators are needed to turn powerful flows of water into useful amounts of electricity. Constructing and maintaining the power lines that carry the electricity to customers is also a significant challenge.

tides in the world."[12] Where China's Qingtang River meets the sea, the rising seawater is channeled into a narrow space, creating a 30-foot (9-m) wave known as a tidal bore that travels upriver at 25 miles per hour (40 km/h). All this moving water represents a lot of kinetic energy.

The most common way to harvest the kinetic energy of tides is to build a tidal barrage. This dam-like structure channels water through a series of turbines. As high tide approaches, water flows into the bay, turning the turbines in one direction. As high tide turns to low tide, the water flow reverses. The water returns to the sea, turning the turbines in the other direction. Another version of this concept uses artificial lagoons. These lagoons fill with water at high tide. The water is released at low tide, traveling through turbines as it retreats to the ocean.

New technologies are being invented and tested to capture kinetic energy from tides. These can resemble wind turbines in that they're designed to allow the water to flow past them, turning blades just as wind turns a windmill. One device being tested in the Bay of Fundy carries tidewater through a large doughnut-shaped device with small propellers inside to capture the energy.

Power from Ocean Waves

Ocean waves are created by the friction of wind striking the water's surface. Waves may look as if they are moving water from one place to another, but this is an illusion. In fact, the water remains in mostly the same spot, simply rising up and down in place.

Any surfer who has ever wiped out knows that waves are powerful. Harnessing that power, though, is a little tricky. In 1799, a French engineer named Pierre Girard filed the first **patent** to use the energy generated by waves. He envisioned a device that used the bobbing of ships moored offshore to power heavy machinery using

WORDS IN CONTEXT

patent
An exclusive right to sell a particular design.

planks and fulcrums, like giant seesaws, to power pumps on land. Although a good idea in principle, the machine was never built, and the devices inventors created in the years following did not generate enough electricity to be practical for large-scale energy production.

In 1974, Scottish inventor Stephen Salter created a device called a nodding duck, a teardrop-shaped floater attached to a vertical spine in the water. The up-and-down motion of the waves would cause the floater to rise and fall. This motion would pump pistons that would push a liquid through a turbine to generate electricity. Other devices used similar floaters attached to pistons anchored to the seafloor. Unable to get major funding, the duck did not enter widespread use. However, it inspired later developments in wave energy. Salter noted that his invention was ahead of its time: "Our only mistake was doing it too early—but that is better than too late."[13]

Power from Ocean Currents

Another way ocean water moves is through currents. An ocean current is like an underwater river, with a large amount of water flowing through an area of ocean that is otherwise mostly still. Currents form as a result of temperature differences. Warm water molecules have more energy and move faster, making them less dense and lighter, while cold water molecules have less energy, slow down, and become denser and heavier. Warm water rises while cold water sinks. The Sun is the primary mover of ocean currents, because it warms water near the equator more than it does near the poles. Cold ocean water at the poles also releases its salt as it freezes, making the leftover brine even heavier as it drops to the bottom of the ocean. Water moves to and from these areas of differing temperatures across Earth.

Ocean currents can be huge. The Gulf Stream, off the southern coast of Newfoundland, discharges 33 million tons (30 million metric t) of water per second. By comparison, all of the rivers around the world combined discharge just 1.1 million tons (1 million metric t) of water per second. The US government's Bureau of Ocean Energy Management says, "It has been estimated that taking just 1/1000th the available energy from the Gulf Stream would supply Florida with 35 percent of its electrical needs."[14] However, this is a potential that is largely untapped. Power plants that use current turbines are currently experimental or on the drawing board.

Humanity may have been getting power from water for millennia, but the way we use this renewable resource is still evolving. Tidal, ocean current, and ocean wave power technology are under active development, and few commercial power plants exist to take advantage of these resources. "Anything on the ocean is difficult.

Researchers have mapped the flow of major currents. These currents circulate water throughout the world's oceans.

It's an unforgiving environment. Part of the problem is that the ocean is rougher because the energy is denser," said Klaus Lackner, professor of sustainable engineering at Arizona State University.[15] Hydroelectric dams, on the other hand, are in widespread use throughout the world. Scientists and engineers are experimenting with different ways to generate energy more efficiently, to store it until it can be used, and to extract energy with as little impact on the surrounding environment as possible. Hydro and wave power have already reduced our dependence on fossil fuels. As technology improves, these promising energy sources will continue to produce more and more clean energy in the future.

CAN HYDRO AND *WAVE* ENERGY REPLACE FOSSIL FUELS?

As useful as hydroelectricity is, in 2014, only 2.4 percent of the world's energy was produced by hydroelectric power plants. By comparison, 81.1 percent of the world's energy was produced by oil, coal, or natural gas. In the United States, these fuels accounted for 81 percent of the country's energy production in 2016. This is a huge portion, though the US Energy Information Administration also noted that it was "the lowest fossil fuel share in the past century."[16] About 10.5 percent of the nation's energy came from renewable sources, and 8.5 percent came from nuclear power.

In the United States, electricity generation makes up 39 percent of energy use, transportation makes up 29 percent, industry makes up 22 percent, and heating homes and businesses makes up 10 percent. These different uses involve varying amounts of each energy source, depending on what works best in each case. For example,

The transportation sector is dominated by vehicles that burn fossil fuels for energy. However, an increasing number of cars run on electricity, and their batteries can be charged with hydroelectric power.

oil provides 92 percent of the energy used in the transportation sector, but only 1 percent of the energy used to generate electricity. Hydroelectric power plants generated 6.1 percent of US electricity in 2015.

Oil (also known as petroleum), coal, and natural gas are all fossil fuels. These are also known as hydrocarbons, since they are made up of hydrogen and carbon atoms chained together. Hydrocarbons can be gases like natural gas, liquids like oil, or solids like coal. Introducing oxygen and a heat source to a fossil fuel results in a reaction known as combustion. It produces carbon dioxide, water, and a lot of energy.

Fossil fuels get their name from the way they were created. They originally existed as plants and animals that lived millions of years ago. When these organisms died, their fossilized remains were covered over by sediment, soil, and rock. Under pressure for millions of years, these materials transformed into hydrocarbons. Fossil fuels allow us to harness the energy that existed in ancient plants and animals.

The Problem with Fossil Fuels

There are two major problems with fossil fuels. One is that fossil fuels are a non-renewable resource. The sources of petroleum, coal, and natural gas we've used since the 1700s took millions of years to produce, and once they're used up, they are gone forever. Humanity has been using fossil fuels at an increasing rate. While experts disagree on how much fossil fuel remains to be extracted, the fact remains that once these reserves disappear, anything that relies on fossil fuels for energy will be useless. Leslie Magoon, a professor at Stanford University's Department of Geological and Environmental Sciences, said in a report to the US Geological Survey in the year 2000, "Technology is great, but it can't find what's not there. In the last five years, we consumed 27 billion barrels of oil a year, but the oil industry discovered only 3 billion barrels a year. So only one barrel was replaced for every nine we used!"[17]

Even if we had an unlimited supply of fossil fuels, another problem with this source of energy is how it disrupts Earth's carbon cycle. Certain gases, including carbon dioxide, are known as **greenhouse gases**.

WORDS IN CONTEXT

greenhouse gases
Gases in the atmosphere that allow the Sun's heat to enter but prevent it from bouncing off Earth's surface and returning to space, like the glass on a greenhouse.

When they are released into the atmosphere, they trap the Sun's heat, warming the planet. This is called the greenhouse effect. Some warming is helpful for the planet, and even necessary for life as we know it. Without greenhouse gases in the atmosphere, the average temperature of the planet would be around 0°F (-18°C). But too much of this effect can quickly warm the planet. Climate scientist Michael Mann says, "We know that the earth is warming up, that we can't explain that warming through natural factors. We can only explain it when we include the effect of increasing greenhouse gas concentrations from human activity."[18]

Carbon dioxide naturally enters the atmosphere when animals breathe. They inhale oxygen and exhale carbon dioxide. Volcanoes and wildfires also release carbon dioxide. Plants absorb carbon dioxide through photosynthesis and combine it with the Sun's energy to sustain themselves and produce oxygen. Animals that eat plants take in this carbon and energy. Plants and animals release carbon dioxide back into the atmosphere when they decompose after death. This movement of carbon dioxide between plants, animals, and the atmosphere is part of the carbon cycle.

Fossil fuels represent millions of tons of plant and animal remains that have been buried under rock for millions of years, removing their carbon from Earth's carbon cycle. Burning fossil fuels over the past few centuries has put back carbon into the atmosphere. As a result of heightened levels of greenhouse gases, the greenhouse effect has intensified. This has led to rising average global temperatures and changes to our climate that will increase unless we can find a way to limit greenhouse gas emissions.

Why Is It So Hard to Stop Using Fossil Fuels?

If the problems of relying on fossil fuels are so obvious, why haven't we switched to clean, renewable resources? While renewable energy sources don't run out and don't release carbon dioxide or other pollutants, there are a number of hurdles renewable energy sources, including hydro and wave power, have to overcome.

Since the 1700s, humanity has relied on fossil fuels because they were easy to find and easy to use. Only small amounts of coal, oil, or natural gas were needed to provide useful energy. Burning coal powered steam engines that could pull long, heavy trains. Natural gas can be passed through pipes and safely burned to heat homes or cook meals. Oil is refined into gasoline, which can power cars and trucks. According to the University of Michigan's Transportation Research Institute, the average car in the United States can travel 25.4 miles (40.9 km) per gallon of gasoline. Considering how much a car weighs, that's a lot of energy packed inside a small space.

Coal, natural gas, and oil are also portable. Coal is a brown or black rock-like substance until it is burned. Oil and its by-products, as well as natural gas, can be kept in containers and moved to wherever they need to be used. Hydroelectric energy is usually used close to the energy supply, either a waterfall or a dam. Transforming hydro power into electricity allows the energy to be transferred far away using wires, but this still won't offset the fossil fuels used in transportation. An electric car can't easily be plugged into its source of electricity as it drives around. Its energy must be stored in batteries, which are expensive and heavy. The vast majority of today's cars run on gasoline.

Finally, there are the political and financial costs of changing the world's economies from being dependent on fossil fuels to using renewable energy. Fossil fuels have been the largest source of energy for our world for decades, and much of our infrastructure has been built with fossil fuels in mind. Battery-powered electric cars exist, but charging stations are relatively rare compared with gas stations. Transforming the infrastructure to allow for the widespread use of electric cars will be an expensive initial cost to overcome. One person seeking to do this has been entrepreneur Elon Musk. The head of electric car company Tesla Motors, he has developed a nationwide network of car charging stations. "I traveled with my family from Los Angeles to South Dakota, to Mount Rushmore using the Supercharger network," Musk noted in 2014.[19] Since that time, more stations have been added to the network.

There is also political resistance to the idea of phasing out fossil fuels. The fossil fuel companies in 2013 made $331 billion in profits in the United States and Canada. In 2014, they employed at least 2 million Americans. People who depend on fossil fuels for their employment are less likely to support political measures to favor renewable resource development. Such resistance has led some politicians to question the science behind global climate change.

Electric car chargers are now found in many cities. Drivers can pay a small fee to recharge their cars. The cost is typically much lower than for a tank of gasoline.

During the bitterly cold winter in February 2015, Oklahoma senator James Inhofe held up a snowball while addressing the US Senate and said, "In case we have forgotten because we keep hearing that 2014 has been the warmest year on record. I asked the chair, do you know what this is? It's a snowball just from outside here. So, it's very, very cold out. Very unseasonable."[20] Critics argued that some politicians with close ties to the fossil fuel industry were intentionally downplaying the dangers of climate change. Writer Phil Plait noted that "oil and gas companies donated nearly half a million dollars to [Inhofe's] campaigns from 2011–2016."[21]

Today, the marketplace is changing as new technology reduces the cost and increases the efficiency of renewable energy. As the number of people who make their living in renewable energy industries increases, the political strength of those industries is likely to increase, too. In 2016, 9.8 million people globally were employed in the renewable energy sector. As this number gets larger, switching from fossil fuels to renewables like hydro power will become politically easier to achieve.

The Advantages of Hydro and Wave Power

Because they rely on gravity and temperature differences caused by the Sun, energy sources from rushing rivers, shifting tides, ocean waves, and underwater currents do not add greenhouse gases or other pollutants to the atmosphere as they generate electricity. And compared with other renewable energy sources, hydro and wave power are relatively reliable. Wind or solar energy can be interrupted by a calm day or cloud cover, but unless there is a severe drought, water remains available.

Hydroelectric power is relatively cheap, too. The National Hydropower Association explains, "States that get the majority of their electricity from hydropower like Idaho, Washington, and Oregon on average have energy bills that are lower than the rest of the country.

Relying only on the power of moving water, hydro prices don't depend on unpredictable changes in fuel costs."[22]

All of these factors make hydro and wave power a strong, environmentally friendly alternative to fossil fuels. Hydroelectric power production already plays a large part in the generation of electricity, producing more than 40 percent of the electricity generated from renewable sources.

Social and Ecological Impacts

Hydro and wave power are not without their problems. One issue is that the power is available only where there is flowing water. One reason coal quickly came to power the United Kingdom during the Industrial Revolution instead of hydroelectricity was due to the country's relatively flat landscape. In many cases, if there is no sharp drop in elevation within a river, power plant builders have to make their own by building hydroelectric dams.

Although hydro power is inexpensive to generate once a dam and turbine are installed, building that dam and installing the turbines is costly. Building dams can require vast amounts of concrete and steel, along with an army of workers. The materials and labor cost millions of dollars, but the environmental cost is significant as well. Transporting workers and materials to the site in vehicles powered by fossil fuels results in enormous amounts of emissions.

The water behind a dam can also have an environmental impact. Hydroelectric dams generate their power by stopping the flow of water

Constructing a dam is a resource-intensive task. Though the hydroelectric power it eventually generates is a clean energy source, making the dam itself releases significant emissions.

of a river and allowing it to build up into a large artificial lake behind the dam. The depth of this lake determines how much power the dam can generate. The deeper the lake, the more of the land surrounding the river must be flooded. Artificial lakes like Abiquiu Lake behind the Abiquiu Dam in New Mexico provide many benefits, such as recreation areas, irrigation for farms, and flood control for communities downstream. However, people living by the river upstream have no choice but to move as water levels rise. Lake Shasta in California flooded eight towns during the construction of Shasta Dam in 1944; the largest of these, Kennett, previously had a population of more than ten thousand.

Making new lakes where none had been before changes the local environment. Where herds of animals once had a narrow river to cross on their migration, they now have a deep lake. New organisms may arrive to take advantage of the suddenly still waters, including mosquitoes, which can bring disease. Plants submerged by artificial lakes can rot, especially in warm climates, producing carbon dioxide and methane that is released into the atmosphere.

Hydroelectric dams can dramatically affect animals within the dammed river itself. Fish such as salmon and steelhead have a life cycle in which they migrate from the ocean into rivers and streams to spawn. Dams block adult fish from their breeding grounds, and baby fish can be sucked into the dam's intake vents and turbines as they head downstream. On March 19, 1934, the US Congress passed the Fish and Wildlife Coordination Act, requiring government agencies to consider the needs of fish and wildlife when building new hydroelectric dams. One design solution was to create fish ladders, passages over the dam where water flows in a series of small waterfalls that fish can swim up or down. Even with these precautions, hydroelectric dams in the Pacific Northwest started to have a serious impact on salmon stocks by the 1940s.

Another problem hydro dams have to deal with is silt. Flowing water causes erosion of the ground it travels over, and the materials the water picks up are taken downstream. When dams stop the flow of water, the sediment settles out of the water and falls to the bottom of the reservoir. If the river that flows into a reservoir has a lot of sediment, a lot can collect behind the dam, making the reservoir shallower. This shrinks the height of the head of the dam and reduces the power the dam can generate. If silt is not removed, some artificial lakes can completely silt over, rendering their dams useless. Researchers estimated that "more than 120 reservoirs [in California] had less than 25 percent of their original capacity remaining and nearly 190 had lost more than 50 percent of their water space."[23]

Dangers from Dams

On July 24, 2010, the Delhi Dam in Iowa saw 10 inches (25.4 cm) of rain fall in twelve hours, overfilling the lake behind the dam. When one of the three spill gates failed to open, the water overtopped the dam, causing the dam to breach and release the lake behind it. Fortunately, officials saw the problem developing and evacuated hundreds of people living downstream. More than fifty homes and twenty businesses were damaged.

One of the worst dam failures of the twentieth century occurred in Morbi, India, on August 11, 1979, when heavy rains overwhelmed the spillways of the Macchu Dam and released all its water on the people downstream without warning. The total number of deaths is unknown, but it is estimated at between five thousand and ten thousand. The floodwaters also swept the topsoil off the farms downstream, reducing the productivity of the future crops. Author Tom Wooten described the disaster this way: "It's like a bathtub with a clogged drain and the faucet on—eventually water would wash over the top. Unfortunately, the top of this bathtub was made of dirt."[24]

People have been generating hydroelectric power for more than a century, so its benefits and risks are well known. Tidal, wave, and current power are newer and more experimental, so many of their challenges are just being discovered. Some can be serious. Most tidal power is generated by building a barrier called a barrage across the mouth of a tidal basin. This disrupts the movement of sea life, which often relies on tidal flats for its food supply or its spawning grounds. Ocean current power works by placing a turbine within an ocean current. Marine animals use currents as part of their migration, and turbines can catch these animals, just as wind turbines sometimes strike birds out of the air. Marine life can also get caught inside machinery, harming the creature and damaging the equipment.

Other Problems

Another problem that hydro, tidal, wave, and current power faces is cavitation. This occurs when high-speed rotors in a turbine form bubbles. The spinning rotors change the pressure of the water passing through, which causes bubbles or cavities to form where the water pressure is low. As the pressure increases again, the bubbles collapse, creating small but intense shock waves that batter at anything the water comes against. These can dent and chip the metal rotors of a turbine, break seals, and poke holes in pipes, destroying the turbine.

The salt water of the ocean is also more corrosive than fresh water. This means that people have to work harder to maintain ocean turbines in good working order. Such maintenance work is a challenge when these facilities are out at sea. And trying to solve these issues can create new problems. As writer Renee Cho notes, "Chemicals used in anti-corrosion coatings or grease for the machinery can leak into ocean waters."[25]

Finally, there is the question of how to transport the power generated by hydro and wave power plants. Many potential hydro and wave power sites are located far from towns and cities. On land, accessing that power means building roads and transmission towers into those areas, disrupting local wildlife. At sea, power cables face corrosion and can disrupt marine life.

Is There Enough Potential Hydro and Wave Energy to Replace Fossil Fuels?

Hydroelectric power plants already have a capacity of more than 100,000 MW in the United States. That is a lot, but it is dwarfed by the amount of energy that Americans consume. Fortunately, there is already a lot of potential hydro and wave energy that has not yet been used. The United States has 2,500 hydroelectric dams, but more than 80,000 dams that do not produce power. Instead, they are used to control flooding or provide water for agriculture. Some of these can be converted to add electrical power to the grid without increasing their ecological impact. There are many other rivers where hydroelectric power plants could be installed.

The potential power available from tidal power, wave power, and ocean current power dwarfs the amount of hydroelectric power potential. As researchers experiment with and improve these new

The construction of Quebec's James Bay project proved highly controversial due to its impacts on First Nations peoples. Government cooperation with the Grand Council of the Cree helped to resolve some of these issues.

technologies, much more green energy can be added to the electric grid. Scientists and engineers in other industries are working on ways to use these forms of energy. Companies like Tesla and General Motors are working on electric cars that they hope will eventually be as inexpensive and easy to fuel as gasoline-powered cars. If they become widespread, electric cars with rechargeable batteries can offset large amounts of fossil fuel use in the transportation sector.

This cooperation with other industries is crucial if hydro and wave power are to replace fossil fuels. Still, these sources are unlikely to provide all of our energy in the future. More likely, they will make up one part of a broad renewable energy portfolio, working alongside such sources as wind and solar to power our world. In eliminating our dependence on fossil fuels, we will likely have to turn to a diverse selection of renewable power to meet our energy needs.

Other Ways of Harnessing the Waves

It may be difficult to have hydro power drive cars, but hydro power can power boats. The key is in the outboard motor, which in many boats is a loud and polluting gasoline engine that powers a propeller in the back of the boat. Sailboats use these motors if the wind isn't available. More environmentally friendly motors use electric power stored in a battery.

However, just as electricity can produce the kinetic energy to turn the propeller's blades, the blades, if moved, can generate electricity that can be stored in the battery. When the wind is blowing above 6 knots (11 km/h), the propeller blades can turn as the boat moves through the water, generating electricity that can be used later when the wind dies down. Such a device is available for small sailboats today.

CHAPTER 3

HYDRO AND *WAVE* ENERGY IN ACTION

ydro energy is one of the oldest forms of renewable energy, but with tidal, wave, and ocean current energy under active development, it is also a field with a lot of innovation. In more than a century of modern hydropower technology, there have been several significant installations that are landmark achievements in this field. They demonstrate the benefits and challenges of hydropower, as well as the scientific, technical, and political forces that have influenced the industry.

Niagara Falls: Power to the People

Niagara Falls was one of the first hydropower resources to be used to generate large amounts of electricity. It started when German-American business mogul Jacob F. Schoellkopf purchased a canal near the falls, expanded it, and installed turbines. By 1882, he was powering seven mills along the New York side of the river. On the Canadian side, a 2,200-kilowatt power plant was built in 1893.

The hydroelectric facilities at Niagara Falls came online in the late 1800s. These power plants provided electricity to both Canada and the United States.

But Niagara Falls was also a natural wonder, and people traveled to the area from miles around to see it. By 1900, a number of private businesses were lobbying to expand power generation from the falls, and the Ontario government knew it had to balance the public's desire to retain the natural beauty of the falls with the benefits of electrical generation. Even among those who wanted to expand Niagara's power generation for Ontario, there was concern that Niagara's power might be controlled by one or two large American companies charging high rates. This issue was fought in the provincial election of 1905. The Conservative Party, led by James Whitney, used the slogan, "The water power of Niagara should be as free as the air."[26] The Conservatives won the election, and Whitney appointed

Adam Beck, a former mayor of London, Ontario, as chairman of the Ontario Hydro Electric Power Commission. Beck set about building transmission lines to distribute Niagara's power across southern Ontario.

Ontario's electric network expanded through the first part of the twentieth century. The demands for power increased, and more hydro power plants were added, including the Abitibi Canyon Generating Station hydropower plant. The power stations drawing energy from Niagara Falls were expanded. On August 15, 1950, the Queenston Power Plant was renamed the Sir Adam Beck Generating Station #1, and work began on Sir Adam Beck Generating Station #2.

Similar things were happening on the American side. The New York Power Authority, led by Robert Moses, sought to replace some of its older power plants on the Niagara River with a large modern power station. With all of the competing demands on Niagara Falls, Canada and the United States negotiated the Niagara Diversion Treaty, signed in 1950. As part of the treaty, the International Control Dam was built to measure the amount of water going over the falls. Both parties had to ensure that a minimum of 748,052 gallons (2,832,000 L) flowed over the falls every second before any water was used for power generation. This ensured that tourists could still come see the beauty of the waterfalls. The agreements between Canada and the United States remain in place, even as both sides increase the efficiency of their power generation capabilities.

The power stations of Niagara Falls were launched when the environmental impacts of hydro power weren't fully understood, but also when governments were more willing to control private businesses, limiting their development of what the government and the public saw as an important natural resource. The current management of Niagara Falls balances power generation with environmental and tourism concerns. This has been achieved through decades of investigation, learning, and negotiation.

Megaprojects: Hoover Dam, James Bay, and Three Gorges

Niagara Falls helped establish the precedent of hydroelectric power plants being projects of national importance. Of the ten largest power-producing facilities in the world today, nine are hydroelectric. They power millions of homes and provide thousands of jobs. Many projects have also affected the environment and disrupted many lives. Three enormous dam projects show the promise and perils of hydroelectricity: Hoover Dam in the United States, the James Bay Project in northern Quebec, and the Three Gorges Dam in China.

Hoover Dam, originally called Boulder Dam, has a generating capacity of 2,080 MW. From 1939 until 1949, it was the largest power plant in the world. It blocks the Colorado River, creating Lake Mead, the largest reservoir in the United States by volume. Although planned in the 1920s, construction work began during the Great Depression in 1931, bringing ten thousand to twenty thousand unemployed people to the site to compete for a few thousand available jobs. When it opened in 1936, President Franklin D. Roosevelt celebrated the dam's contribution to the American economy, and the spirit the engineers and workers had shown over nature itself:

> In a little over two years this great national work has accomplished much. . . . To employ workers and materials when private employment has failed is to translate into great national possessions the energy that otherwise would be wasted. Boulder Dam is a splendid symbol of that principle. The mighty waters of the Colorado were running unused to the sea. Today we translate them into a great national possession.[27]

The James Bay Project was launched in the early 1970s by the government of Quebec, led by Premier Robert Bourassa. From 1972 to 1985, workers constructed several hydro power stations in northern Quebec with a total generating capacity of 10,300 MW. A workforce of up to twelve thousand employees moved 203 million cubic yards (155 million cubic m) of fill and used massive amounts of concrete and steel. A second phase, carried out between 1987 and 1996, added another 5,200 MW of generating capacity.

Premier Bourassa saw the James Bay project as vital in bolstering Quebec's economy and energy independence. At a speech given on April 30, 1971, he said, "The development of James Bay is the key to Quebec's economic progress, it is also the key to social progress and political stability: it is the future of Quebec."[28]

The Three Gorges Dam is a gigantic installation by any measure. However, it also attracted controversy because its reservoir flooded cities and historical sites.

The Three Gorges Dam in China is the largest hydro dam in the world. Opened in 2003 after many years of construction, it has a generating capacity of 22,500 MW. The Three Gorges Dam blocks the Yangzi River. The dam is 7,661 feet (2,335 m) long, and 594 feet (181 m) high. At the official launch of construction in 1997, Chinese president Jiang Zemin said that the "remarkable feat in the history of mankind to reshape and exploit natural resources . . . embodies the great industrious and dauntless spirit of the Chinese nation."[29] All three projects transformed the countries that built them. However, these projects also resulted in large social and environmental problems.

The Environmental Impacts of Big Dams

The first and biggest issue around large hydroelectric dams is the land they flood behind them. The reservoir behind the Three Gorges Dam covers 419 square miles (1,085 sq km). Lake Mead behind Hoover Dam covers 247 square miles (640 sq km). The James Bay Project flooded eight reservoirs totaling 4,440 square miles (11,500 sq km) of wilderness. This flooding has been blamed for changes in the migratory routes of herd animals, including the drowning of more than seven thousand caribou in northern Quebec on one weekend in 1984.

In James Bay, the flooding of boreal forests released mercury, a toxic element that had been present in the vegetation, into the water supply. The Cree First Nations people who hunted and fished in the area saw mercury levels rise in their population in the years that followed. Mercury is poisonous, with exposure linked to brain development problems and learning disabilities. Environmental monitoring has helped limit exposure.

An important difference between Niagara Falls and the Hoover Dam and James Bay projects was that Niagara Falls was closer to the communities it powered. It was already known as a natural wonder, and people cared strongly about its environmental well-being. The more remote Hoover Dam and James Bay projects took longer to generate environmental concerns from the public. For many, these sites were out of sight and out of mind.

The Social Impacts of Big Dams

Big dams also cause significant social impacts. Dam construction employs thousands of people, often bringing them to remote locations where they have to be housed and fed. Then there are the people affected by the flooding after the dam is built. Over the years, different governments have responded to these challenges in different ways.

Lake Mead's remote location in the arid southwestern United States limited this disruption, but one town, Saint Thomas, Nevada, was flooded out. The town had a peak population of five hundred. As the waters rose, the residents tore down their homes and sold their land to the government. The last citizen to leave was Hugh Lord, who rowed away from his home on June 11, 1938. Other dam projects farther upstream on the Colorado River brought the US government into conflict with the Navajo Nation, which campaigned to protect sacred sites from flooding and overuse by tourists. The National Park Service has set up a plan for consulting the local Native American nations on the management of their sacred sites, but these disputes continue.

The James Bay project pitted the Quebec government against the Cree Nation of northern Quebec, who saw little benefit from the project and whose hunting lands were flooded and polluted. The Cree protested through the 1970s and 1980s, especially as the Quebec government planned to dam the Great Whale River. As people in North America became more concerned about the environment in the late 1980s, the leaders of the Cree Nation were able to gain international attention and sympathy for their cause. In 1992, the New York government withdrew from a multi-billion-dollar agreement to purchase power from Quebec, citing a decrease in energy requirements and the outcry over the Great Whale project.

The construction of Quebec's James Bay project proved highly controversial due to its impacts on First Nations peoples. Government cooperation with the Grand Council of the Cree helped to resolve some of these issues.

The Quebec government and the Grand Council of the Cree were able to negotiate an agreement over the James Bay project called *La Paix des Braves*. The agreement gave the Cree joint jurisdiction over the Quebec lands around James Bay and promised the Cree would benefit more from the jobs and revenues raised by the hydroelectric project. In return, the Grand Council of the Cree agreed to further hydroelectric development of two more rivers in northern Quebec, adding 1,368 MW of additional generating capacity. The grand chief of the Cree, Ted Moses, approved of Quebec premier Bernard Landry's decision to negotiate this settlement: "He understands that the Crees must be part of Quebec's vibrant economy and a living part of its economic and cultural mosaic."[30]

China's Three Gorges Dam is even more controversial. The Chinese government argues that the dam's power reduces the country's coal use by enough to reduce greenhouse gas emissions by 110 million tons (100 million metric t). However, the Three Gorges Dam's reservoir flooded 244 square miles (632 sq km) of land, including thirteen cities, 140 towns, and more than 1,600 villages, forcing the relocation of 1.3 million people. The Chinese government has claimed that the dispossessed people received payment, new homes, and jobs, but people have complained that funds have not reached their intended recipients. The Chinese government has been criticized for using excessive force to quell protests over the resettlement problems.

A report by the International Rivers Network concluded, "The lack of independent grievance mechanisms and the punishment meted out against peaceful protesters violate China's own laws. [The Chinese government is] in breach of the country's obligations under international law, including the Covenant on Civil and Political Rights, which China has signed. The widespread human rights violations also present a challenge to the governments which fund the Three Gorges Project through official export credits and guarantees."[31]

According to the World Commission on Dams, large dams have displaced as many as 40 million people from their lands over the past sixty years. There have been further costs, including losing access to clean water and food supplies. Because hydropower projects can be so disruptive, it is important for their builders to engage with the local communities. Minimizing the social impacts of the dam, maximizing efficiency, and reducing the environmental effects are all key parts of modern hydropower.

Tidal Power Projects

The first power plant able to develop industrial amounts of electricity from tidal power was the Rance Tidal Power Station in France. Built between 1961 and 1966, it uses a 2,460-foot (750-m) barrage to capture the tidal power from the Rance River Estuary. Its twenty-four turbines can generate up to 240 MW of electricity.

Rance was one of only two large-scale tidal power stations in the world until 1980. The other was in Russia. The number of tidal estuaries that can generate large amounts of power is limited, and concerns exist over the environmental impact of Rance's barrage. Critics suggest that it blocks marine life, and they have noted that sand eels and other fish have disappeared from the estuary since the power plant was installed. By taking power from the tides, the barrage reduces the movement of water in the estuary, resulting in more silt being deposited within the local **ecosystem**. The company operating the power plant is trying to limit the plant's environmental impact.

More recently, the Sihwa Lake Tidal Power Station opened in South Korea in 2011. Generating

WORDS IN CONTEXT

ecosystem
An environment and its community of organisms that interact with each other. When one part is affected by some change, other parts of the ecosystem are affected.

up to 254 MW of electricity, it passed Rance as the largest tidal power station in the world. The station also uses a tidal barrage, converting a seawall that had been built in 1994 to reduce tidal flooding. The seawall itself had been an environmental problem, allowing pollutants to build up in the reservoir behind it. The Sihwa Lake power station was seen as a means of reintroducing seawater into the reservoir at high tide. It moves water around and cleans things up while generating electricity at the same time. Countries and companies continue to research new technologies that will allow power plants to reap the power of the tides with less impact on the ecosystem.

Wave Power Projects

Wave power technology is even more experimental. Many **prototypes** try to harvest energy by anchoring part of a power generator to the ocean floor while allowing another part to float on the surface, rising and falling against the anchor to generate power.

One such model came to be known as Salter's Duck. This was an experimental system designed in 1974 by University of Edinburgh professor Stephen Salter to convert wave power to energy. At the time, the world was going through an oil crisis. Middle Eastern oil-producing countries raised the price of oil by 300 percent, sparking a boom in the research for alternative energy supplies.

WORDS IN CONTEXT

prototype
A test machine or model built to test the effectiveness of a design, from which other machines are developed.

Salter's prototype floated a 20-foot- (6-m-) long spine on the ocean. Mounted on this spine were twelve pear-shaped objects, like duck heads, that pointed in the direction of incoming waves. As a wave **oscillated**, it would cause these ducks to rock backward on the spine. This energy would be transferred to pumps, which would push liquid through a turbine to generate electricity.

> **WORDS IN CONTEXT**
>
> **oscillate**
> To rise and fall, or to shift back and forth between states regularly, as in a wave.

Salter envisioned floating ducks the size of houses. It was estimated that a single full-scale duck could generate 6 MW of electricity, enough to power four thousand homes. This enthusiasm was tempered by the fact that this electricity cost $1 per kilowatt-hour, more expensive than other sources, but Salter hoped that further tests could improve the design and make it cheaper and more efficient.

In the early 1980s, oil prices dropped sharply. The research program funding Salter was shut down by the British government. At a meeting of the Committee on Science and Technology in the British parliament in 2001, Salter submitted a memo saying, "If I had to supply reasons for the failure of the first UK wave programme, I would cite over-optimism, the attempt to make very big (2GW) power stations and to assess infant devices too quickly. The programme was properly supported and enthusiastically led from 1976 to 1983, a period of only seven years, and then entered a very unhappy phase where researchers felt that they were always on the defensive."[32]

In 2008, the Scottish company Pelamis Wave Power reached an agreement with the Portuguese government to establish a wave farm 3 miles (5 km) offshore in the Portuguese parish of Aguçadoura. The farm consisted of three 400-foot (120-m) floating cylinders, each split into four separate sections that rocked and shifted with the waves. As each segment moved in relation to the others, the motion pumped oil under high pressure through turbines.

The Pelamis machines were able to generate electricity, but technical glitches made this prototype unreliable, and within six months, the project was shut down. Bearings in the machines did not stand up to the pounding of the waves, and the mechanisms had to be redesigned. Although a solution was found, money to fix the problem dried up. An economic downturn hurt the Australian power company that was sponsoring the project, and so the experiment was never restarted.

But this wasn't the end of wave power. That same year, the government of Gibraltar announced a partnership with the Israeli-based company Eco Wave Power to install a wave farm off its coast. Using a design they tested off the coast of Jaffa in Israel, Eco Wave Power installed a series of floaters anchored to an old **jetty**

WORDS IN CONTEXT

jetty

A stone, metal, or wooden structure that juts out from land into water to provide a mooring space for ships and to protect a harbor from incoming waves.

that rise and fall as the waves move beneath them, pumping pistons that push liquid through turbines. Unlike the Portuguese experiment, which had a peak capacity of 2.25 MW, the first phase of this wave plant generates just 100 kilowatts of energy. However, it is hoped that by 2020, when other wave generators are installed, the Gibraltar plant

SeaGen and Ocean Current Power

Commercial ocean current power projects do not yet exist. The concept remains experimental, although the United States and the United Kingdom researched possible sites in 1974 in response to the oil crisis. These studies suggested the potential power generation was huge, but commercial developers couldn't overcome the costs of building these offshore generators and connecting them to the electrical grid. Questions also remained over such generators' impact on sea life.

In the early 1990s, in response to climate change concerns, the British government created Marine Current Turbines Ltd., which set up a turbine that was sunk into Loch Linnhe on the coast of Scotland. The turbine harvested 15 kilowatts of energy from the movement of the tides. It also proved that the turbine could work capturing power from current. Based on this, in 1998, a larger project called Seaflow was built and installed in the same area, generating 300 kilowatts from 2003 until it was decommissioned and removed in 2007. An even larger current turbine called SeaGen was built in 2008 and installed in the Strangford Narrows in Northern Ireland. It's rated at 1.2 MW for all currents flowing at 5.4 miles per hour (2.4 m/s) and faster.

SeaGen faced a number of challenges, including construction on the ocean floor and fixing problems that occur out at sea. There were concerns about SeaGen's impact on marine life, but as the rotors turned at a speed of only fourteen revolutions per minute, seals and fish easily avoided being hit. In general, SeaGen met its power generation goals, and the technology is being considered for future installations in the United Kingdom and Canada.

will have a generation capacity of 5 MW, enough to meet 15 percent of Gibraltar's needs. In the words of Gibraltar's chief minister Fabian Picardo, "At last we are seeing the grid fed with renewable energy, something that was long overdue. . . . But it's even more exciting than that. This is the first time, in the whole of Europe, that a renewable wave energy system is linked into an electricity grid. The Gibraltarians are here pioneers in our partnership with Eco Wave Power."[33]

Eco Wave Power has already launched follow-up projects, including a plan to build a 4.1 MW wave power plant at the Port of Manzanillo in Mexico, as well as another plant in the Zhejiang Province of China. In its press release, Eco Wave Power said, "With over [11,000 miles (18,000 km)] of coastline and approximately 6500 islands, China is believed to be one of the biggest markets for wave energy. The theoretical mean power of wave energy resources along China's coasts reaches 12852.2 MW, with the most massive resources in Taiwan, Zhejiang, Guangdong, Fujian, and Shandong."[34]

The Major Players in Hydro and Wave Power

Since the construction of Niagara Falls, Hoover Dam, and the James Bay project, other countries in the world have developed their hydroelectric potential to a greater degree. Of the ten largest hydro power plants in the world, four are in China, and only one (the Grand Coulee Dam) is in the United States.

China is the most populous country in the world and has a rapidly growing and modernizing economy. This growth has led to severe pollution from fossil fuels, and China hopes that moving to renewables such as solar and hydro power will lead to a more sustainable and environmentally friendly future.

Three of the five largest power plants in the world, all of them hydro power, are found in South America. Brazil operates the Tucuruí Dam and shares the Itaipu Dam with Paraguay. Venezuela operates the Guri Dam across its Caroní River. Colombia is also working on the Ituango Dam, which could generate 2,400 MW of electricity when it opens in 2018. In Asia, Pakistan, Myanmar, Tajikistan, and India are building dams able to generate 2,000 MW of power or more. In Africa, Angola and Nigeria are working on projects that will generate more than 5,000 MW. Ethiopia is damming the Blue Nile with the Grand Ethiopian Renaissance Dam, which could generate more than 6,000 MW of power, making it the largest hydroelectric power plant in Africa. The name itself describes the hopes the government of Ethiopia has for the project.

The United States and Canada are major players because they have large rivers to exploit. Hydro and wave power can be generated only where there is water. Similarly, the major players in tidal powers are fortunate to have areas that experience major tides. France was the pioneer in tidal energy with its Rance tidal power plant, and South Korea's Sihwa Lake station is the world's largest. Nova Scotia's Bay of Fundy offers a tremendous amount of kinetic energy from its tides, and the province's Annapolis Royal Generating Station is one of the older tidal power stations in the world, having opened in 1984. Nova Scotia Power is testing an experimental device that could generate power more efficiently and with less impact on marine life. England and Scotland have built two tidal energy stations generating 326 MW of power. In total, nineteen countries are investing in research into tidal power.

Wave and ocean current energy are making progress, too. This field is seeing private companies in Scotland, the United States, Portugal, and elsewhere experiment and innovate. David Ferris, a journalist specializing in energy and climate, writes, "The endeavor of turning ocean energy into usable electricity is experiencing a Wild West of innovation."[35] Hydro and wave power offer exciting opportunities for large- and small-scale developments, and well-established and experimental technologies.

The Challenges of Hydro and Wave Energy

Big hydro and wave projects can generate large amounts of green and renewable electricity. However, governments and project managers need to be aware that these projects come with social and environmental costs, not only for people living in the flood plains behind the dams, but for people and wildlife living downstream, people and wildlife who depend on tidal estuaries, and plants and animals in the ocean. Projects need to be carefully designed to limit problems and ensure all benefits are distributed fairly. Governments must be ready to listen and address the concerns of everybody affected. Through planning and negotiation, these problems can be addressed, but it takes patience and political will.

*W*HAT IS THE FUTURE OF HYDRO AND *W*AVE ENERGY?

The scientific consensus is that greenhouse gases are causing global climate change and the human race needs to cut back on burning fossil fuels in order to prevent climate change from getting worse. Even without global climate change, the human race cannot continue to use fossil fuels at the present rate forever. These fuels will run out eventually, so humanity will have no choice but to transition to renewable resources. There is a bright future for hydro and wave energy if we can produce this energy with less impact on the environment, at lower cost, and in more places.

In many cases, advances in technology lead to bigger and more powerful things. Some of the largest hydro power plants in the world are being built in China, and China has led the way in installing some of the largest turbines in the world. The 6.4-gigawatt Xiangjiaba Dam features eight 800-MW Francis turbines. Managing such a huge dam

This aerial view shows Chile's Rapel Hydroelectric Plant. The dams of hydroelectric power plants hold back enormous amounts of water. These power stations produce a large portion of today's renewable energy.

requires extensive control and safety systems. Computers monitor water pressure, generation, and vibrations, sending that information to control centers many miles away. Brazil hopes to open the Belo Monte dam in 2019. It could generate more than 11,000 MW of energy. Luiz Augusto Barroso, a Brazilian government official in the country's energy planning agency, notes that Brazil has a huge untapped potential for hydroelectric power: "In the developed world almost 70 percent of the hydro potential has already been exploited, whereas here in Brazil, 70 percent of our hydro has not been explored yet."[36]

Finding Opportunities to Grab Energy

One key to creating green energy is to generate it efficiently. A common way to do this is to find places where people could be generating electrical power but aren't. Potential hydroelectric power can be made at thousands of dams in the United States that do not currently produce power. Converting these into hydro power plants can add to the electrical power of the grid with a minimal increase to their impact on the environment.

This isn't limited just to dams. The city of Portland, Oregon, generated hydroelectric power from its own water pipes when a company called LucidEnergy installed turbines that were turned by water flowing through the city's drinking supply. The turbines didn't affect the water in any way, did not affect any wildlife any more than the pipes did, and generated enough electricity to power 150 homes. These turbines not only catch and harvest energy that would otherwise be wasted, but also have the added benefit of allowing Portland to monitor its water supply, detecting drops in pressure and alerting work crews to possible burst pipes. These turbines can be used in any pipe where water is pulled by the force of gravity, including sewer pipes and wastewater pipes. On its website, LucidEnergy describes problems faced by water utilities and then discusses their devices as a solution:

> *The high cost of energy, coupled with energy efficiency mandates and the need to repair or replace aging infrastructure all require creative solutions to keep operations sustainable. By using their water pipelines to generate renewable energy from an otherwise untapped energy source, the LucidPipe Power System can be part of the solution.*[37]

This could even be applied at the micro scale, in the pipes within a home. As a person turns on the water to brush his or her teeth, a little generator could be generating electricity.

New Innovations

Then there are innovations that could increase the amount of electricity produced by the power plants already in place. The penstock pipe, invented by engineers David Piesold and Colin Caro, has spiral ridges along its sides. These ridges cause the water to spin within the pipe as it flows. This allows the power plant to focus more of the water's force against the blades of the turbine, boosting the power produced by as much as 10 percent.

Another way to increase the efficiency of hydro and wave power plants is to pair them with other forms of renewable energy. Strong winds often blow at the tops of cliffs and atop dams, and the ocean has wide open spaces where the wind blows freely. Installing wind turbines or solar panels alongside hydro and wave power plants can boost energy production. The facilities can also share the infrastructure required to send power to the grid, reducing the overall impact on the environment.

The pairing of renewable power plants is not the only way that different forms of renewable energy can help the development of hydro and wave energy. Innovations in wind energy have been applied to possible tidal and ocean current power generation, and vice versa. The ocean current project SeaGen experimented with a new rotor blade design that proved more efficient in turning the kinetic energy of moving water into electrical power. The new design was based on the latest designs from wind turbines. Although the principle behind wind turbines and current power generators is the same, the fact that

A SeaGen turbine underwent testing in 2010. The experimental turbine was slated for decommissioning in 2017, but the lessons learned during testing would be applied to future designs.

water has more force than air meant that SeaGen's rotors had to be built much tougher, using carbon fiber for additional strength. Martin Wright, the director of the company behind SeaGen, said of the project, "I hope it makes people believe that tidal power isn't 20 to 30 years away and a dream, but it is something that, if we get the right resources around it, could become a significant reality and contributor much quicker than that."[38]

Scientists and engineers have spent a lot of time researching ways to make machines run more efficiently. If turbines can be turned with less force, we can generate more energy with less work. New materials can make turbines lighter and stronger, and new lubricants can help reduce friction so that less energy is wasted as heat.

Some researchers have even considered ways of generating usable electricity from friction itself. Scientists at the Georgia Institute of Technology are investigating how the triboelectric effect generates electricity when two materials move against each other, such as when a person gets an electric shock walking across a carpet and touching a metal doorknob. A prototype generator created by the research team could harvest some of the wasted energy of friction in this manner. Not only could this improve the efficiency of turbines, but it could be a way to directly grab energy from ocean currents and ocean waves without using turbines or tidal barrages. Professor Zhong Lin Wang said, "We are able to deliver small amounts of portable power for today's mobile and sensor applications. This opens up a source of energy by harvesting power from activities of all kinds."[39]

The Advantages of Going Small

Small, efficient turbines can be used to increase the number of locations where hydro power can be useful. Small-scale hydroelectric projects such as small dams, wave-energy turbines, or simple waterwheels may not feed a lot of power into the grid, but they may not need to. They can provide enough power to light up a remotely located town or village in the developing world.

This can even be done on the personal level. Plans already exist on the Internet for individuals to build small hydroelectric generators. Using a bucket to hold water, PVC pipes to control the flow, and a small turbine to generate power as the water flows out, people can generate enough power to charge cell phones and personal electronics wherever there is water to fill a bucket. Designed by Sam Redfield, this system provides cheap, clean, renewable energy without using batteries or solar panels, which is good not only for camping but

The Hydroelectric Sneaker

In 1999, Canadian inventor Robert Komarechka filed for a patent titled "footwear with hydroelectric generator assembly." His concept features a sneaker with two sacs of liquid in the sole, with a connection between them. As the wearer walks, the sacs are squeezed differently, causing liquid to flow back and forth between them. A tiny generator placed in the connector produces power from this movement. In terms of hydroelectric power, the shoe design works in a similar way to a tidal power station, capturing the changing direction of the liquid as force is acted on it by different parts of the foot.

"United States Patent 6,239,501," United States Patent and Trademark Office, May 29, 2001. www.patft.uspto.gov.

also for communities in the developing world that don't have access to an electrical grid. These small generators are already providing power to small clusters of homes in Peru and Guatemala, and unlike solar power generators, these mini-hydro power plants work twenty-four hours a day. The project's website describes it as "a low-cost, appropriate-technology solution to the problem of supplying electricity to those at the bottom of the economic ladder."[40]

WORDS IN CONTEXT

decentralized
When something is produced or managed in several places within a system.

A **decentralized** power grid using many smaller power plants located closer to users rather than fewer big power plants located far away is more stable, and it doesn't require long transmission lines to carry electricity to its users. And small-scale waterwheels powering a mill or a home is how hydroelectric power got started, millennia ago.

Power Storage

As society builds toward a decentralized power grid, with more small-scale solar and wind power installations, a key to making this system work will be devising better ways to store large amounts of energy until it is needed. One possibility is to store water's kinetic energy as potential energy. This is done at pumped storage plants. At these facilities, pumps send water uphill into a storage basin or natural lake. This increases the water's potential energy. Then, when the water leaves the storage basin and heads back down to the level where it was pumped, it is channeled through turbines and generates electricity. Sometimes the turbines themselves can act as pumps, reversing direction when needed.

Because energy is needed to power the pumps, pumped-storage plants do not generate more power than it takes to run them. However, these plants are useful for storing energy, like batteries. At certain times, electricity from the grid is cheaper, and the pumps send energy uphill. Later, when electrical demand rises, the water is released to generate electricity.

Pumped storage capacity is seen as a way to increase the usefulness of all renewable energy sources. The National Hydropower Association says, "The more grid energy storage we have—using tested technologies like pumped storage—the more new energy resources we can bring online."[41]

Considering the Implications

Whatever new power is generated from hydro or wave energy, care must be taken to ensure that each development is built in a way that helps, rather than hurts, the environment, and that provides benefit not only for energy users far away, but for people at the site of the power plant who may be affected by its construction and use. Large-scale power plants such as those in China have been criticized for hurting people in the area affected by rising floodwaters, changing the local environment, impacting the local food supply, polluting, and more.

Researchers are looking at ways of mitigating the problems associated with hydroelectric dams and other hydro and wave power projects. Fish ladders have been used to allow salmon and other fish to get down and upstream past hydroelectric dams as they go through their reproductive cycle. The Thompson Falls hydroelectric plant in Montana took this further, building a 72-foot- (22-m-) high fish ladder with forty-eight pools where the fish could rest. At the top of the ladder is a 17-foot- (5-m-) long gathering pool and holding tank where researchers examine and tag the fish as they pass through, allowing scientists to study the dam's impact on the fish life more closely. A US Fish and Wildlife official described the project as a win for all parties involved: "The Thompson Falls project is a good example of state, federal, tribal and a private utility partnership to provide natural resource protection."[42]

Instead of fish ladders, one company suggested sending salmon over a dam using a fish cannon. The company, named Whooshh Innovations, uses flexible tubes and vacuum pressure to safely and gently suck fish from the river below the dam, depositing them in the reservoir above. Prototypes are being tested by the US Department of Energy and the Washington Department of Fish and Wildlife. The company describes its system as an "efficient, cost-effective, environmentally sustainable and better way to move live and harvested fish safely, gently, and hygienically."[43]

Hydro and Wave Power Flow Forward

Hydro and wave power look to the past and to the future. Hydro energy was one of the first renewable energy sources, and the technology humans perfected over millennia helped them more easily establish electrical grids during the late nineteenth century. As the International Hydropower Association points out, "It was hydropower that set [England's] industrial revolution running. In many regions of the world, hydropower has played an equally major role in increasing and transforming development."[44] At the same time, new technologies and techniques make it possible to extract power from water more efficiently, and to obtain power from water in places we wouldn't have thought possible decades ago. Experiments with tidal power, wave energy, and ocean current power have shown that the available energy from moving water is huge. The combination of tried-and-true technology and innovative inventions makes these sources of energy some of today's most exciting fields.

INTRODUCTION: ENERGY FROM WATER

1. Quoted in Kevin Bissett, "Could Bay of Fundy Tides Generate Enough Power for All of Atlantic Canada?" *CBC News*, May 30, 2016. www.cbc.ca.

2. Quoted in Chris Wood, "Gibraltar's Landmark Wave Power Station Opens for Business," *New Atlas*, May 26, 2016. www.newatlas.com.

3. Quoted in "Worlds First Large-Scale Tidal Energy Farm Launches in Scotland," *Guardian*, September 12, 2016. www.theguardian.com.

4. Quoted in Jason Margolis, "Wave Farms Show Energy Potential," *BBC News*, March 2, 2007. www.bbc.com/news.

5. "Environmental Impacts of Hydroelectric Power," *Union of Concerned Scientists*, n.d. www.ucsusa.org.

CHAPTER 1: HOW DO HYDRO AND WAVE ENERGY WORK?

6. Quoted in B. J. Lewis, J. M. Cimbala, and A. M. Wouden, "Major Historical Developments in the Design of Water Wheels and Francis Hydroturbines," *IOP Publishing*, 2014. www.iopscience.iop.org.

7. Quoted in R. A. Gregory, *Discovery: Or The Spirit and Service of Science*. London: Macmillan and Company, 1916, p. 3.

8. "Lowell Waterpower System," *American Society of Civil Engineers*, n.d. www.asce.org.

9. Quoted in "Hydro-electricity Restored to Historic Northumberland Home," *BBC News*, February 27, 2013. www.bbc.com.

10. Quoted in "The Electric Light," *New York Herald*, January 4, 1880, p. 6.

11. Quoted in "A Concrete 'Colossus': The Hoover Dam at 75," *NPR Books*, June 8, 2010. www.npr.org/books.

12. "Tidal Energy Project in the Bay of Fundy," *Government of Canada*, June 26, 2017. www.nrcan.gc.ca.

13. "Ocean Current Energy," *Bureau of Ocean Energy Management*, n.d. www.boem.gov.

14. Quoted in "Inventor of 'Duck' Technology Wins Saltire Prize Medal," *BBC News*, March 23, 2011. www.bbc.com/news.

15. Quoted in Renee Cho, "Tapping into Ocean Power," *Earth Institute*, February 14, 2017. blogs.ei.columbia.edu.

CHAPTER 2: CAN HYDRO AND WAVE ENERGY REPLACE FOSSIL FUELS?

16. "Even as Renewables Increase, Fossil Fuels Continue to Dominate U.S. Energy Mix," *U.S. Energy Information Administration*, n.d. www.eia.gov.

17. Quoted in L. B. Magoon, "Are We Running Out of Oil?" *U.S. Geological Survey*, 2000. www.usgs.gov.

18. Quoted in Richard Moore, "Michael Mann: Man-Made Climate Change Is Real," *Northwoods River News*, April 25, 2016. www.rivernewsonline.com.

19. Quoted in Tom Risen, "Elon Musk Talks Tesla's Supercharger Future," *U.S. News & World Report*, June 11, 2014. www.usnews.com.

20. Quoted in Ted Barrett, "Inhofe Brings Snowball on Senate Floor as Evidence Globe Is Not Warming," *CNN Politics*, February 27, 2015. www.cnn.com/politics.

21. Quoted in Phil Plait, "A Snowball's Chance in Paris," *Slate*, October 27, 2015. www.slate.com.

22. "Why Hydro: Affordable," *National Hydropower Association*, n.d. www.hydro.org.

23. Quoted in George Skelton, "California's Reservoirs Are Filled with Gunk, and It's Crowding Out Room to Store Water," *Los Angeles Times*, March 6, 2017. www.latimes.com.

24. Quoted in Katherine Xue, "Alumni: History Detectives," *Harvard Magazine*, March–April 2012. www.harvardmagazine.com.

25. Quoted in Cho, "Tapping into Ocean Power."

CHAPTER 3: HYDRO AND WAVE ENERGY IN ACTION

26. Quoted in Howard Hampton and Bill Reno, *Public Power: The Fight for Publicly Owned Electricity*. Toronto: Insomniac Press, 2003, p. 37.

27. Quoted in Franklin Delano Roosevelt, "Address at the Dedication of Boulder Dam," *The American Presidency Project*, September 30, 1935. www.presidency.ucsb.edu.

28. Quoted in "Constitution and Subject," *Société de développement de la Baie-James*, n.d. www.sdbj.gouv.qc.ca/en/home.

29. Quoted in Steven Mufson, "The Yangtze Dam: Feat or Folly?" *Washington Post*, November 9, 1997. www.washingtonpost.com.

30. Quoted in Ted Moses, "European Tour 2002: Notes for a Speech by Grand Chief Dr. Ted Moses," *Grand Council of the Crees*, November 2002. www.gcc.ca.

31. Quoted in "Human Rights Dammed Off at Three Gorges: An Investigation of Resettlement and Human Rights Problems in the Three Gorges Dam Project," *International Rivers Network*, January 2003, p. 28. www.internationalrivers.org.

32. Quoted in S. H. Salter, "Memorandum Submitted by Professor S. H. Salter, Department of Mechanical Engineering, University of Edinburgh," *House of Parliament*, February 9, 2001. www.parliament.uk.

33. Quoted in "First Renewable Wave Energy System in Europe Is Launched in Gibraltar," *GBC*, May 26, 2016. www.gbc.gi.

34. Eco Wave Power, "Eco Wave Power Establishes a Subsidiary in Suzhou, and Receives an Approval for a First Plant in Zhejiang Province," *PR Newswire*, March 26, 2015. www.prnewswire.com.

35. Quoted in David Ferris, "What the Future of Wave Energy Looks Like," *Forbes*, September 27, 2012. www.forbes.com.

CHAPTER 4: WHAT IS THE FUTURE OF HYDRO AND WAVE ENERGY?

36. Quoted in "Amazon Culture Clash over Brazil's Dams," *BBC News*, January 10, 2017. www.bbc.com/news.

37. "Case Studies," *LucidEnergy*, n.d. www.lucidenergy.com.

38. Quoted in Alok Jha, "First Tidal Power Turbine Gets Plugged In," *Guardian*, July 17, 2008. www.theguardian.com.

39. Quoted in John Toon, "Harvesting Electricity: Triboelectric Generators Capture Wasted Power," *Georgia Tech News Center*, December 9, 2013. www.news.gatech.edu.

40. "About," *Five Gallon Bucket Hydroelectric Generator*, n.d. www.five-gallon-bucket-hydroelectric.org.

41. "Pumped Storage Strengthens the Grid," *National Hydropower Association*, n.d. www.hydro.org.

42. Quoted in "PPL Montana Dedicates Fish Ladder at Thompson Falls Dam and Hydro Plant," *HydroWorld.com*, September 9, 2010. www.hydroworld.com.

43. "What We Do," *Whooshh Innovations*, n.d. www.whooshh.com.

44. "A Brief History of Hydropower," *International Hydropower Association*, n.d. www.hydropower.org.

BOOKS

L. E. Carmichael, *Hybrid and Electric Vehicles*. Minneapolis, MN: Abdo Publishing, 2013.

Michael A. Hiltzik, *Colossus: Hoover Dam and the Making of the American Century*. New York: Free Press, 2010.

Kathryn Hulick, *Energy Technology*. Minneapolis, MN: Abdo Publishing, 2016.

Renee Corona Kolvet, *Hoover Dam*. Charleston, SC: Arcadia Pub, 2013.

Gregory McNamee, *Careers in Renewable Energy: Your World, Your Future*. Masonville, CO: PixyJack Press, Inc., 2014.

WEBSITES

Bureau of Ocean Energy Management
https://www.boem.gov

This US government website provides information on the exploration of technologies to extract energy from the ocean.

Center for Climate and Energy Solutions
https://www.c2es.org

This site offers information on the environmental impacts of different sources of energy production, including hydroelectric energy.

New York Times: **Hydroelectric Power**
https://www.nytimes.com/topic/subject/hydroelectric-power

This news feed gives updates on the latest stories about hydroelectric power from the *New York Times*. Learn about the companies, scientists, engineers, and politicians who are shaping the future of hydroelectric power.

US Geological Survey
https://www.usgs.gov

This site provides data and information on the geographical and environmental aspects of the United States, with maps, data, discussion, and more.

US Office of Energy Efficiency & Renewable Energy – Clean Energy Jobs and Career Planning
https://energy.gov/eere/education/clean-energy-jobs-and-career-planning

This website from the Department of Energy features information about how to enter a career in alternative and green energy production.

INDEX

IMAGE CREDITS

cover: Constantine Androsoff/ Shutterstock Images

4 (top): Georgios Kollidas/ Shutterstock Images

4 (bottom): Sheila Terry/ Science Source

5: Tupungato/Shutterstock Images

7: TDK Visuals/Shutterstock Images

11: Antoine2K/Shutterstock Images

15: Photo Researchers/ Science Source

23: Red Line Editorial

27: Gary Hincks/Science Source

29: chuyuss/Shutterstock Images

34: Scharfsinn/Shutterstock Images

37: CHAIYA/Shutterstock Images

42: Zimcerla/Shutterstock Images

45: Science Source

49: PRILL/Shutterstock Images

52: Jeffrey Ulbrich/AP Images

63: Andrew J. Billington/ Shutterstock Images

66: Paul Faith/Press Association/PA Wire URN:8512489/AP Images

ABOUT THE AUTHOR

James Bow is the author of more than forty educational books for children and young adults. He is also a novelist and a local columnist. He graduated from the University of Waterloo School of Urban and Regional Planning in 1991. Born in Toronto, he lives in Kitchener, Ontario, with his author wife and his two daughters.